INSECTS OF ALL KINDS

Lynn M. Stone

The Rourke Book Company, Inc.
Vero Beach, Florida 32964

PHOTO CREDITS
© Lynn M. Stone: cover, p. 4, 7, 8, 11, 16;
© J. H. "Pete" Carmichael: title page, p. 12, 13, 19;
© James H. Robinson: p. 15;
© James P Rowan: p. 20

EDITORIAL SERVICES
Janice L. Smith for Penworthy Learning Systems

Library of Congress Cataloging-in-Publication Data

Stone, Lynn M.
    Insects of all kinds / Lynn M. Stone.
        p. cm. — (Six legged world)
    ISBN 1-55916-312-7
    1. Insects—Juvenile literature. [1. Insects.] I. Title.

QL467.2 .S788 2000
595.7—dc21
                                                        00–036926

Printed in the USA

CONTENTS

KINDS OF INSECTS

Naming insects is a big job. Scientists have named about 800,00 **species**, or kinds, of insects. But their work has just begun. There may be another nine or ten million insect species that scientists haven't seen! Many of these "undiscovered" insects are in the world's rain forests. Because rain forests are being rapidly destroyed, many insect species will never be known.

More of the world's named insects belong to the beetle order than to any other group.

INSECT VARIETY

One of the most interesting things about the great number of insect species is their variety. Insects display a remarkable number of shapes, sizes, and colors.

Insects may be shaped liked leaves or twigs or even bird droppings. Some have wide wings or furry bodies. An insect may be a caterpillar one day and a **cocoon** the next.

Insects come in every color you could imagine. Some beetles shine like new metal.

The ability of insects to survive often depends on their strange shapes and amazing colors.

ENTOMOLOGISTS

Entomologists are scientists who study insects. Entomologists place insects into large groups of species called orders. The members of an order share certain likenesses in their body design. Butterflies and moths, for example, both belong to the order called **lepidoptera**.

Most entomologists sort insects into 26 orders. But the greatest number of insects belong to just four of these orders.

An entomologist studies butterflies at a science lab in a Costa Rican rain forest.

THE BIG FOUR

The four orders with the most insect species are the beetles group, the moths and butterflies group, the flies group, and the wasps, bees and ants group.

The beetles have the largest number of known species, about 350,000. Beetles tend to have hard, shiny backs. Several beetle species live in fresh water and a few live in salt water.

*This luna moth belongs to the order **lepidoptera** – the butterflies and moths.*

Flies are part of a large insect order. Flies have only one pair of wings.

The praying mantis and other insects of the mantid order
have large, spiny forelegs for holding prey.

13

There are more than 100,000 kinds of moths and butterflies. Butterflies usually fly by day while moths are active at night. Moths are generally furry and have feathery antennas.

The family of wasps, ants and bees has about 100,000 known species. Many of these insects live together in nests or homes called **colonies**.

Members of the fly family have only one pair of wings. Flies have **mouthparts** designed to gather liquid food.

Bees make up a large order of insects along with their cousins, ants and wasps.

OTHER INSECT GROUPS

Dragonflies and damselflies are among the most beautiful insects. They have slender, brightly colored bodies, see-through wings and huge eyes.

Dragonflies may fly as fast as 60 miles (96 kilometers) per hour. They are swift **predators**, or hunters. Dragonflies use their legs to catch other insects in flight.

The grasshoppers, katydids and crickets make up a family of plant-eaters. Katydids and crickets are known for their night songs.

Dewdrops sparkle on a dragonfly's wings. The dragonfly can't fly until its body warms up.

Walkingsticks and leaf insects look remarkably like their namesakes – sticks and leaves. One walkingstick species is the longest North American insect at seven and one-half inches (178 centimeters).

Another well-known group of insects is the cockroach order. Cockroaches can live in human houses. Most cockroach species live in forests, however.

A grasshopper, shown here, belongs to the same order of insects as katydids and crickets.

The familiar praying mantis is one of the mantid insects. Mantids have large forelegs to hold their **prey**, the animals they catch.

Members of the termite group always live in colonies. Many species build towerlike mounds of mud up to 18 feet (5.5 meters) tall. Some termites build nests with chewed wood. Others build nests with sloped roofs for rain control.

Cicadas, leafhoppers, spittlebugs and aphids belong to an order of insects with beak-like piercing mouthparts. The beaks pierce plants from which the insects suck juices.

The ambush bug, a member of the insect order of bugs, lives on other insects that it catches.

Some people call all insects "bugs". But there actually is an order of insects called bugs. Many bugs are water-loving, like the giant water bug.

Bugs have piercing mouthparts. They use them to suck plant or animal juices. The bed bug, for example, will suck human blood.

GLOSSARY

cocoon (kuh KOON) — a moth's pupa and its silk covering

colonies (KAH luh neez) — groups of animals of the same kind; places where groups of the same kind of animals nest, roost, or raise young

entomologist (en tuh MAH luh jist) — a scientist who studies insects

lepidoptera (leh puh DAHP tuh ruh) — a large order of insects that includes butterflies and moths

mouthparts (MOUTH pahrts) — the jaws, beaks, or other feeding tools that surround an insect's mouth

predator (PRED uh tur) — an animal that hunts and kills other animals for food

prey (PRAY) — an animal that is hunted for food by another animal

species (SPEE sheez) — within a group of closely related animals, such as moths, one certain type (**luna** moth)

FURTHER READING

Find out more about the kinds of Insects and insects in general with these helpful books and information sites:

- Feltwell, John. *Butterflies and Moths*. Dorling Kindersley, 1993
- Goor, Ronald and Selsam, Millicent. *Backyard Insects*. Scholastic, 1981
- Rowan, James P. *Ants*. Rourke, 1993
- Rowan, James P. *Dragonflies*. Rourke, 1993

Wonderful World of Insects on-line at www.insect-world.com
Butterfly Website at butterflywebsite.com
Children's Butterfly Site at www.mesc.nbs.gov
Insects on-line at www.letsfindout.com/bug

INDEX